Shortcuts to College Calculus Refreshment Kit

Shortcuts to College Calculus Refreshment Kit

Juan Acevedo

Copyright © 2012 by Juan Acevedo.

ISBN: Softcover   978-1-4691-6023-8
       Ebook   978-1-4691-6024-5

All rights reserved. No part of this book may be reproduced or transmitted in any form or by any means, electronic or mechanical, including photocopying, recording, or by any information storage and retrieval system, without permission in writing from the copyright owner.

This book was printed in the United States of America.

To order additional copies of this book, contact:
Xlibris Corporation
1-888-795-4274
www.Xlibris.com
Orders@Xlibris.com
111172

Table of Contents

Basic Geometry 1
Quadratic Equation 6
Laws of Exponents 6
Laws of Logarithms 7
Basic Graphs 7
Basic Transformations 12
Basic Trigonometric Functions 15
Parts of a Circle 16
Basic Trigonometric Graphs 17
Basic Trigonometric Identities 20
Sum and Difference Identities 21
Basic Pythagorean Identities 22
Double Angle Identities 22
Basic Trigonometric Transformations 23
Basic Arc Function Graphs 27
Basic Hyperbolic Function Definitions 30
Relationships Between Hyperbolic Functions 31
Basic Hyperbolic Function Graphs 32
Basic Derivatives 35
Basic Integrals 37

Rectangle:

$$\text{Perimeter} = 2x + 2y$$
$$\text{Area} = yx$$

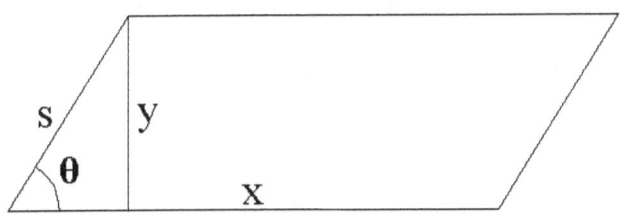

Parallelogram:

$$\text{Perimeter} = 2s + 2x$$
$$\text{Area} = yx = sx\sin\theta$$

Triangle

Perimeter = a + x + b
Area = ½(xy)

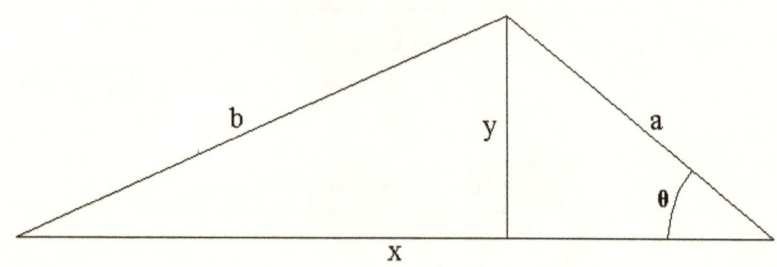

Trapezoid

Perimeter = a + b + h(csc θ + csc φ)
Area = ½ h(a+b)

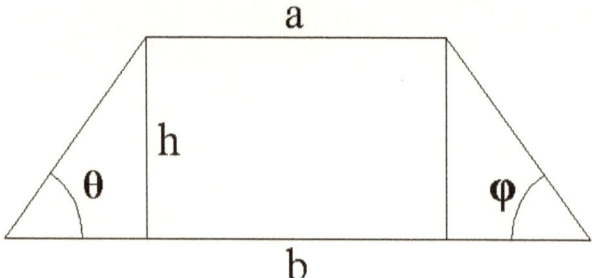

Polygon of n sides

Perimeter = ns
Area = ½ ns² cot π/n

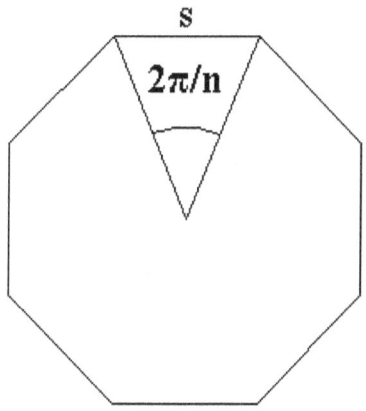

Circle of radius r

Perimeter = 2πr
Area = πr²

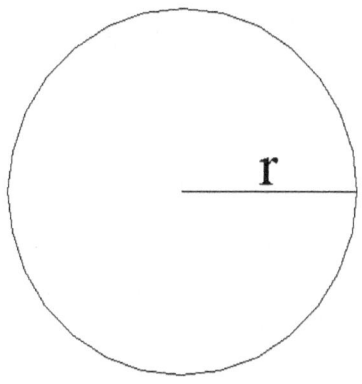

Sector of circle of radius r

Arc Length = s = rθ
Area = ½ (r²θ)
[θ in radians]

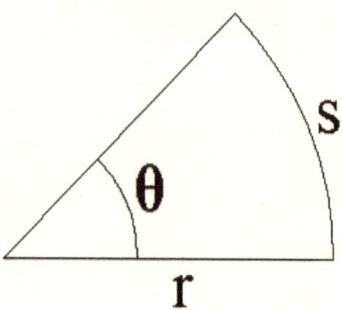

Sphere of radius r

Surface Area = 4πr³
Volume = 4/3 (πr³)

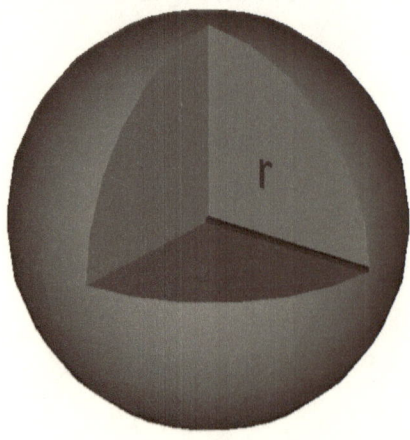

Right circular cylinder of radius r and height h

Lateral Surface Area = $2\pi rh$
Volume = $\pi r^2 h$

Quadratic Equation:
For quadratic equations of the form

$$ax^2 + bx + c$$

$$x = \frac{-b \pm \sqrt{b^2 - 4ac}}{2a}$$

Laws of Exponents

$$x^p * x^q = x^{p+q}$$
$$(x^p)^q = x^{pq}$$
$$(xb)^p = x^p b^p$$
$$x^0 = 1 \text{ when } x \neq 0$$
$$x^p/x^q = x^{p-q}$$
$$x^p = 1/x^p$$
$$\sqrt[n]{x} = x^{1/n}$$
$$\sqrt[n]{x^m} = x^{m/n}$$
$$\sqrt[n]{x/y} = \frac{\sqrt[n]{x}}{\sqrt[n]{y}}$$

Laws of Logarithms

$$\log_a AB = \log_a A + \log_a B$$

$$\log_a A/B = \log_a A - \log_a B$$

$$\log_a A^p = p \log_a$$

$$\frac{\log_b B}{\log_b a} = \log_a B$$

Basic Graphs

$$y = x$$

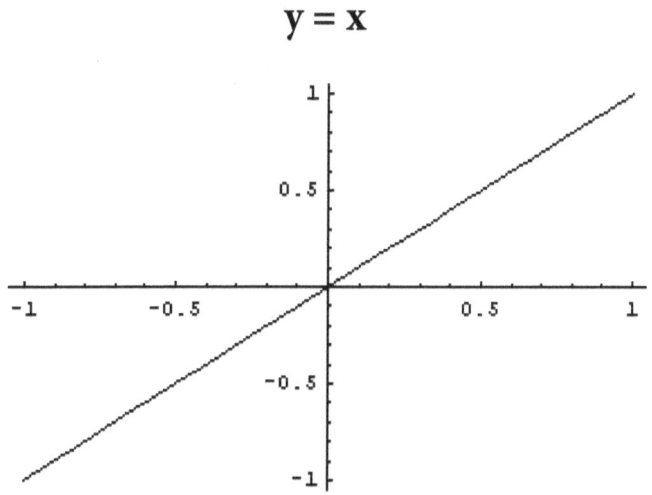

$y = -x$

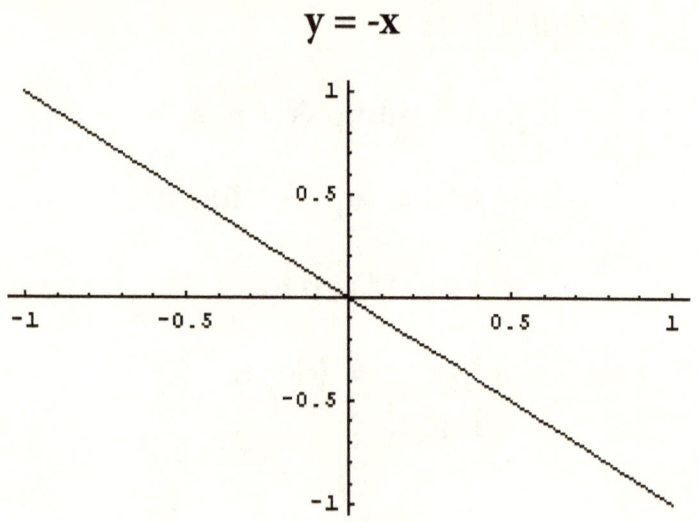

$y = |x|$

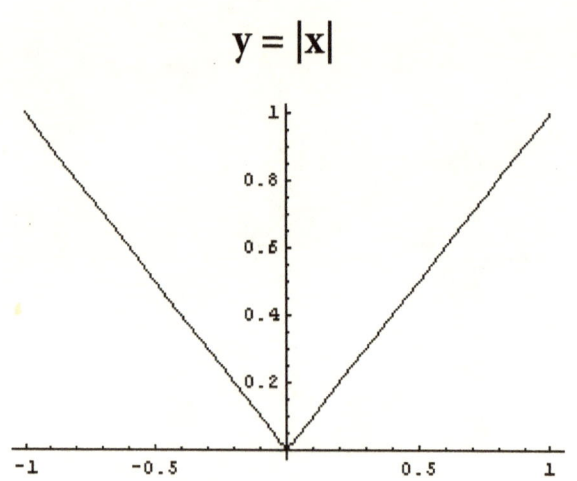

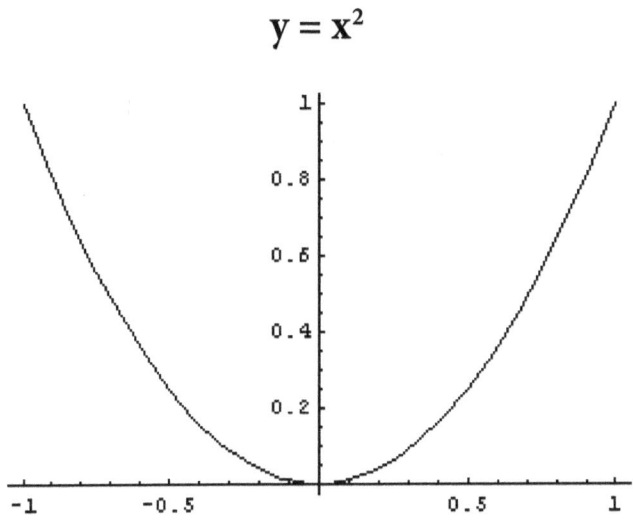

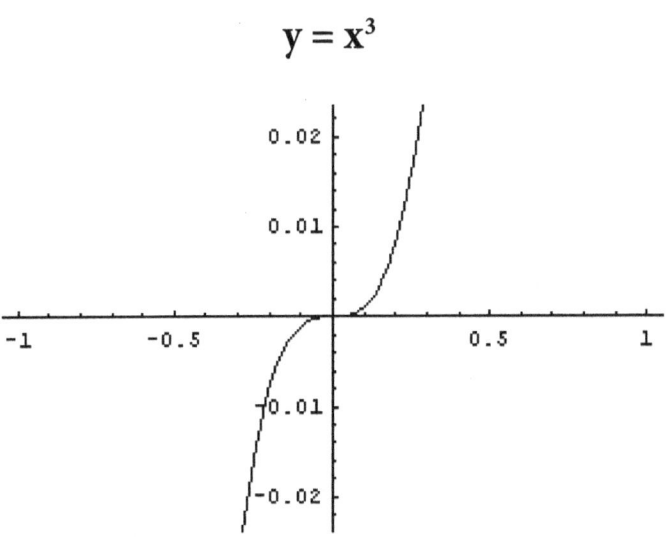

y = 1/x

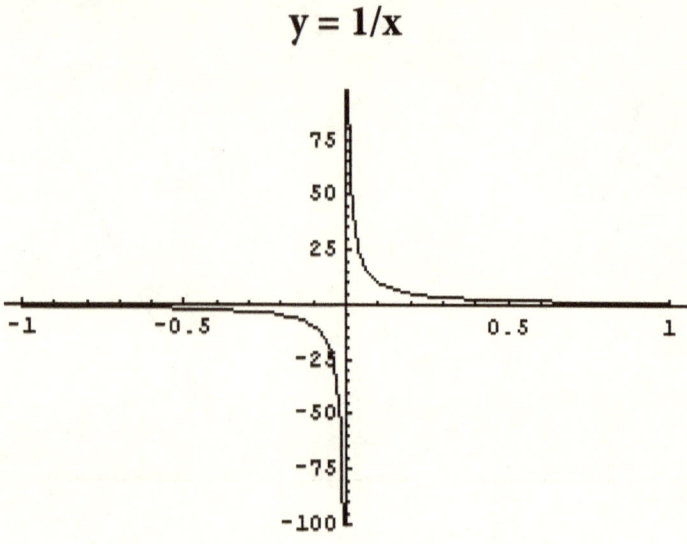

y = 1/x²

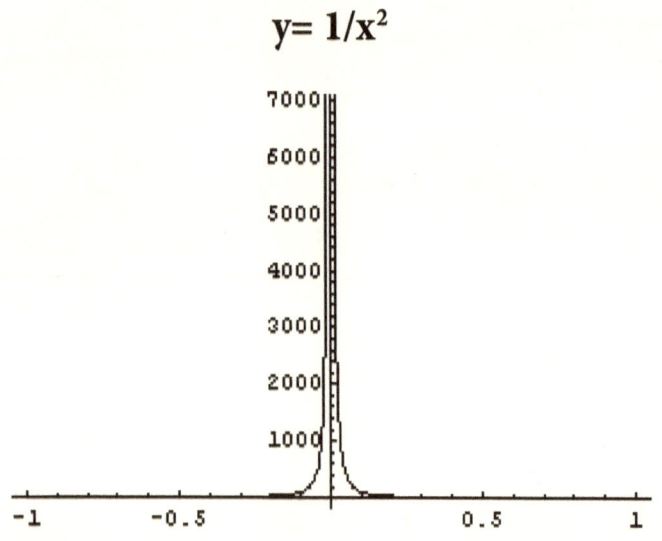

$y = e^x$

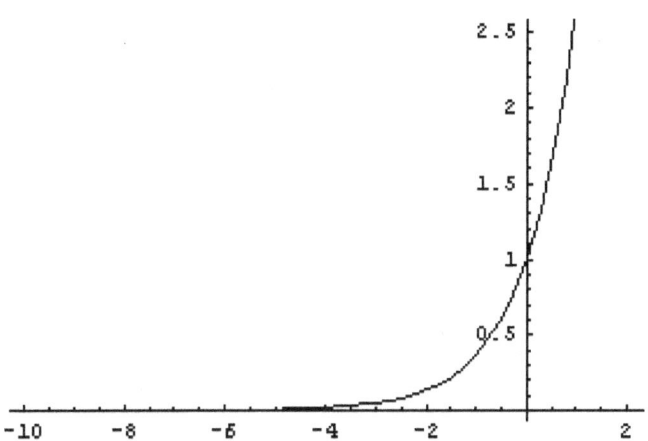

$y = \ln x$

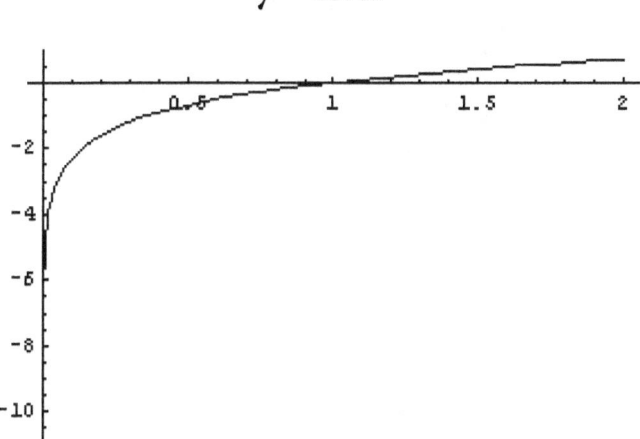

Basic Transformations

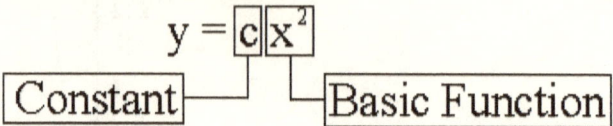

The constant "c" when multiplied by the function x^2 controls the rate at which the basic function increases.

Ex.

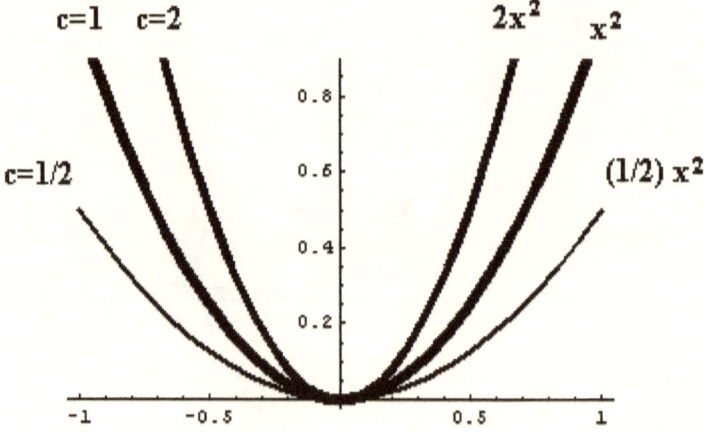

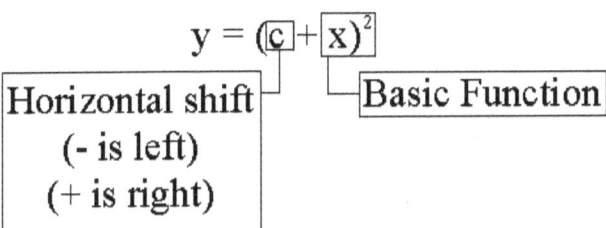

The constant "c" when inside the parenthesis with the function x^2 controls how many units the original graph is shifted to the left or the right.

Ex.

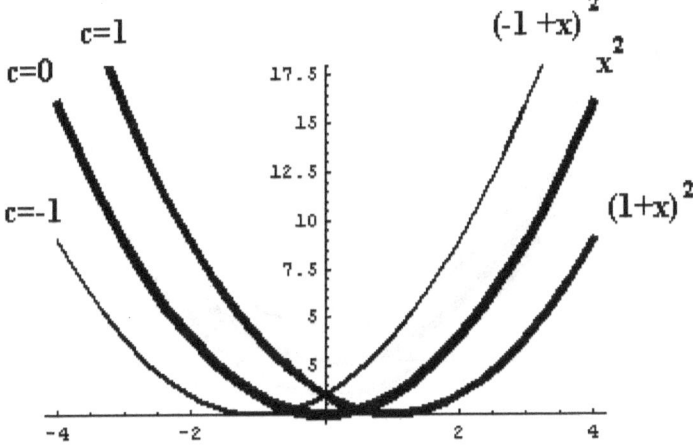

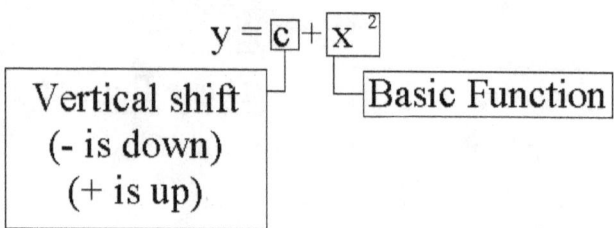

The constant "c" when outside the parenthesis of the function x^2 controls how many units the original graph is shifted up or down.

Ex.

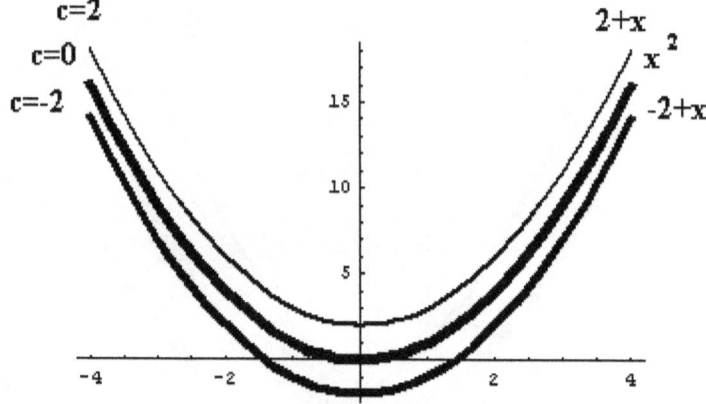

Basic Trigonometric Functions

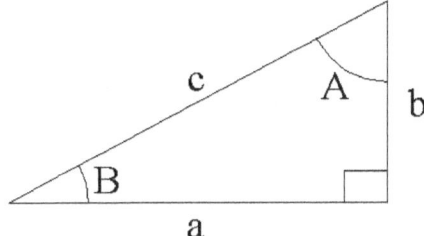

$$\sin(B) = \frac{b}{c} = \frac{\text{opposite}}{\text{hypotenuse}} = \frac{1}{\csc(B)}$$

$$\cos(B) = \frac{a}{c} = \frac{\text{adjacent}}{\text{hypotenuse}} = \frac{1}{\sec(A)}$$

$$\tan(B) = \frac{b}{a} = \frac{\text{opposite}}{\text{adjacent}} = \frac{1}{\cot(B)}$$

$$\csc(B) = \frac{c}{b} = \frac{\text{hypotenuse}}{\text{opposite}} = \frac{1}{\sin(B)}$$

$$\sec(B) = \frac{c}{a} = \frac{\text{hypotenuse}}{\text{adjacent}} = \frac{1}{\cos(B)}$$

$$\cot(B) = \frac{a}{b} = \frac{\text{adjacent}}{\text{opposite}} = \frac{1}{\tan(B)}$$

Parts of a Circle

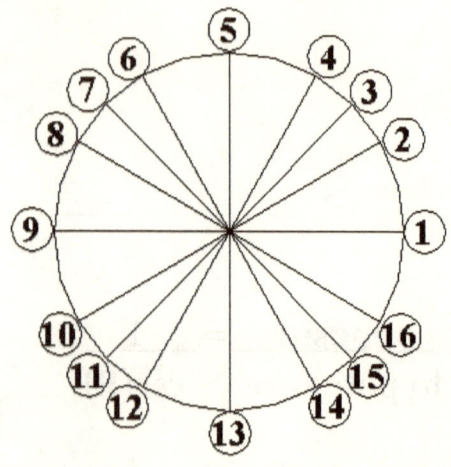

Number	Degrees	Radians	sin(x)	cos(x)	tan(x)
1	0	0	0	1	0
2	30	π/6	1/2	√3/2	√3/3
3	45	π/4	√2/2	√2/2	1
4	60	π/3	√3/2	1/2	√3
5	90	π/2	1	0	Undefined
6	120	2π/3	√3/2	-1/2	-√3
7	135	3π/4	√2/2	√2/2	-1
8	150	5π/6	1/2	-√3/2	-√3/3
9	180	π	0	-1	0
10	210	7π/6	-1/2	-√3/2	√3/3
11	225	5π/4	√2/2	-√2/2	1
12	240	4π/3	-√3/2	-1/2	-√3
13	270	3π/2	-1	0	Undefined
14	300	5π/3	-√3/2	1/2	-√3
15	315	7π/4	-√2/2	√2/2	-1
16	330	11π/6	-1/2	√3/2	-√3/3
1	360	2π	0	1	0

Basic Trigonometric Graphs

sin(x)

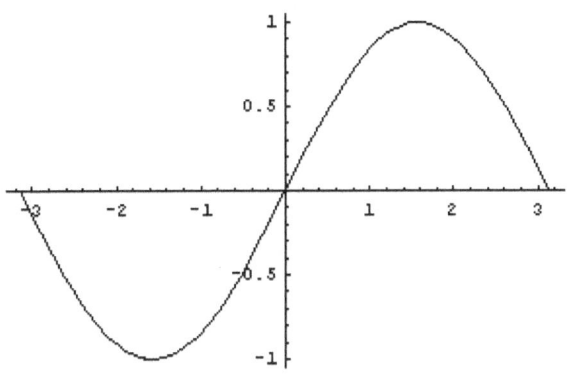

csc(x)

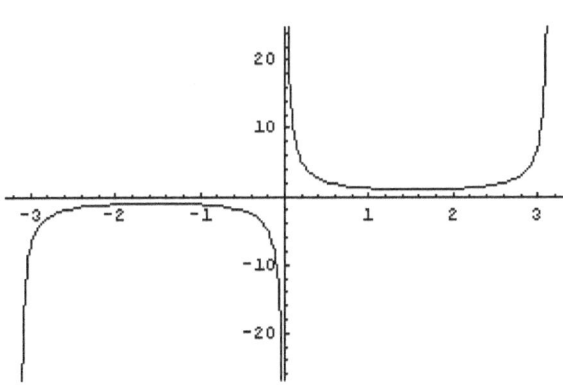

cos(x)

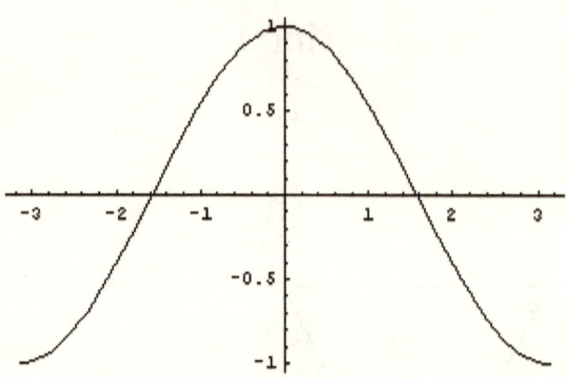

sec(x)

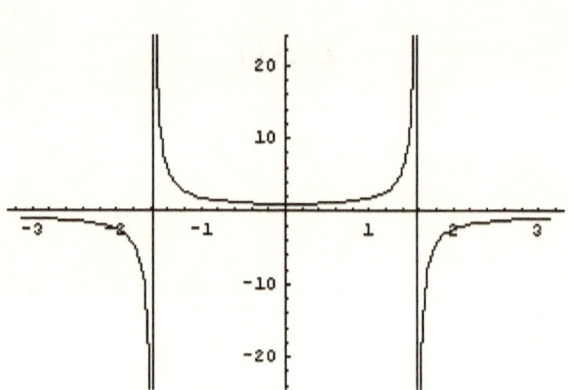

tan(x)

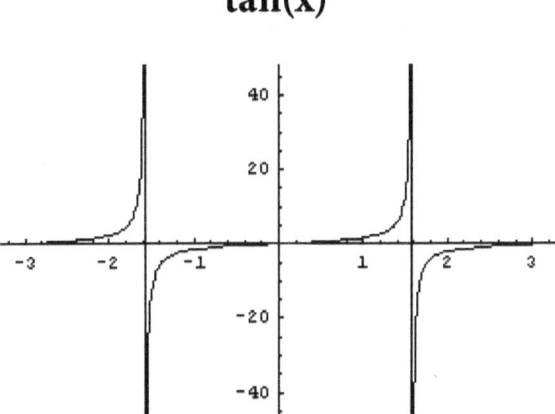

cot(x)

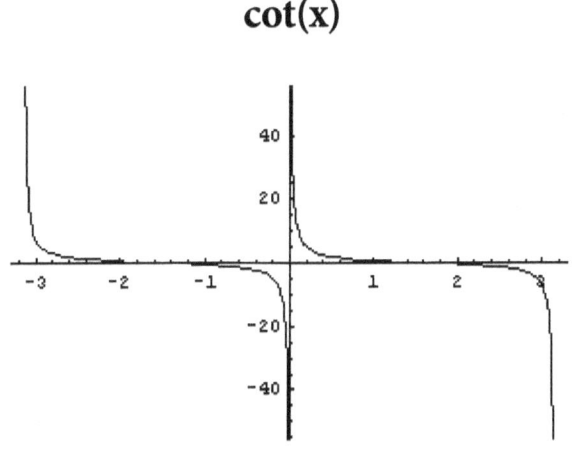

Basic Trigonometric Identities

$\sin(-x) = -\sin(x)$

$\cos(-x) = \cos(x)$

$\tan(-x) = -\tan(x)$

$\tan(x) = \dfrac{\sin(x)}{\cos(x)}$

$\cot(x) = \dfrac{\cos(x)}{\sin(x)}$

$\csc(x) = \dfrac{1}{\sin(x)}$

$\sec(x) = \dfrac{1}{\cos(x)}$

$\cot(x) = \dfrac{1}{\tan(x)}$

Sum and Difference Identities

$\sin(x+y) = \sin(x)\cos(y) + \cos(x)\sin(y)$

$\sin(x-y) = \sin(x)\cos(y) - \cos(x)\sin(y)$

$\cos(x+y) = \cos(x)\cos(y) - \sin(x)\sin(y)$

$\cos(x-y) = \cos(x)\cos(y) + \sin(x)\sin(y)$

$\tan(x+y) = \dfrac{\tan(x) + \tan(y)}{1 - \tan(x)\tan(y)}$

$\tan(x-y) = \dfrac{\tan(x) - \tan(y)}{1 + \tan(x)\tan(y)}$

Basic Pythagorean Identities

$\sin^2(x) + \cos^2(x) = 1$

$\sec^2(x) - \tan^2(x) = 1$

$\csc^2(x) + \cot^2(x) = 1$

Double Angle Identities

$\sin(2x) = 2\sin(x)\cos(x)$

$\cos(2x) = \cos^2(x) - \sin^2(x)$
$\qquad\quad = 1 - 2\sin^2(x)$
$\qquad\quad = 2\cos^2(x) - 1$

$\tan(2x) = \dfrac{2\tan(x)}{1-\tan^2(x)}$

$\sin^2(x) = \dfrac{1 - \cos(2x)}{2}$

$\cos^2(x) = \dfrac{1 + \cos(2x)}{2}$

Basic Trigonometric Transformations

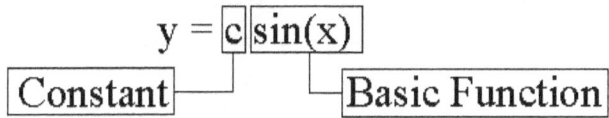

$y = c\sin(x)$ where c is the Constant and $\sin(x)$ is the Basic Function.

The constant c when multiplied by the function sin(x) controls the maximum height of the functions peak and valley.

Ex.

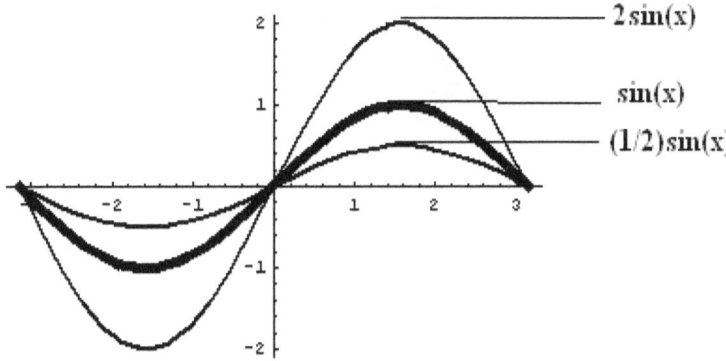

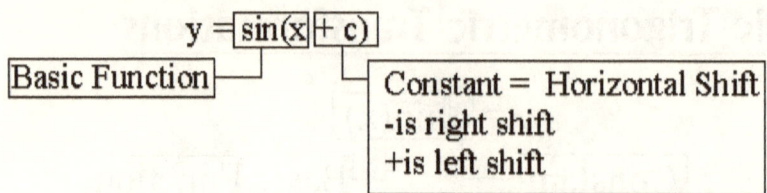

The constant c when inside the parenthesis with the function sin(x) controls how many units to the left or the right the graph is shifted.

Ex.

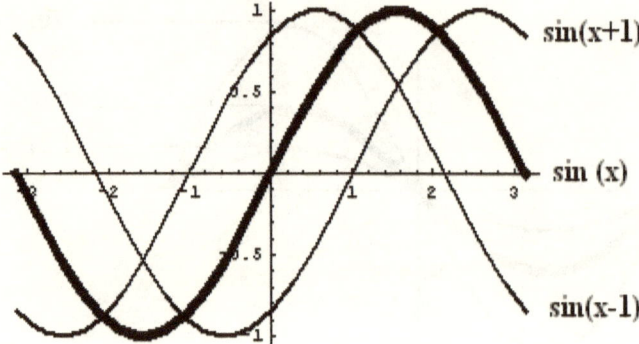

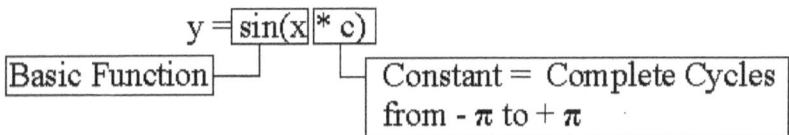

The constant c when multiplied inside the parenthesis of the function sin(x) controls how many complete cycles are shown in one period.
(from $-\pi$ to π)

Ex.

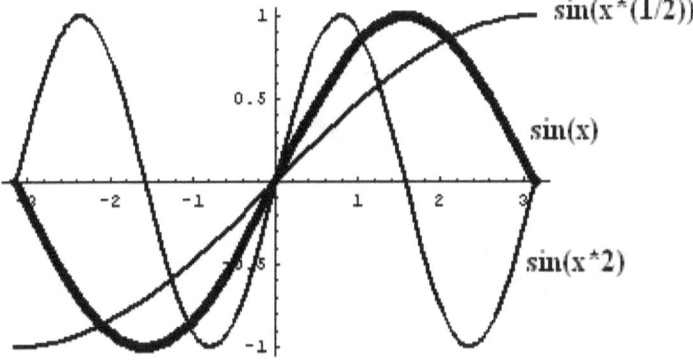

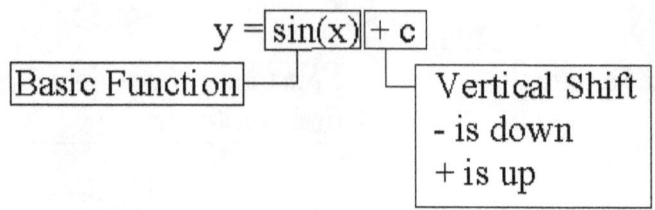

The constant c when outside the parenthesis of the function sin(x) controls how many units the function sin(x) is shifted up or down.

Ex.

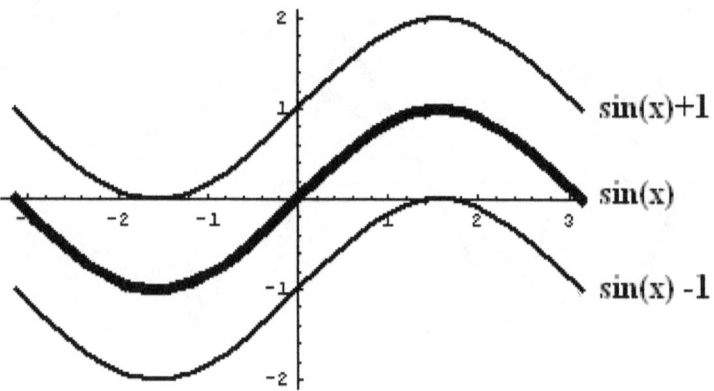

Basic Arc Function Graphs

arcsin(x)

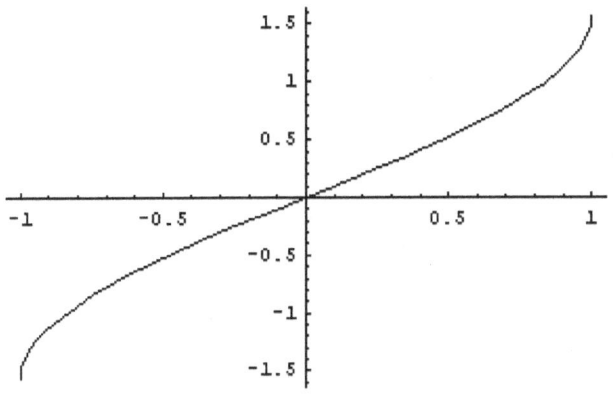

arccsc(x)

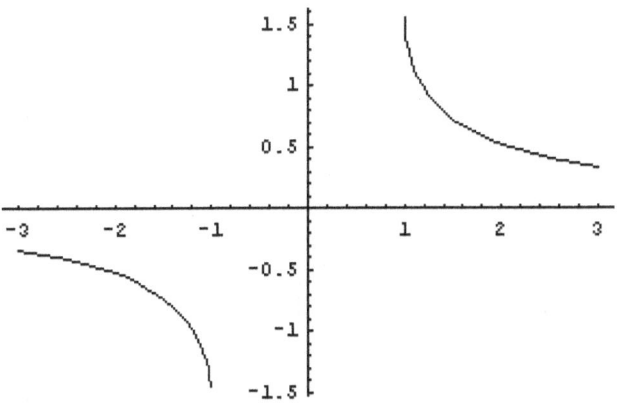

arccos(x)

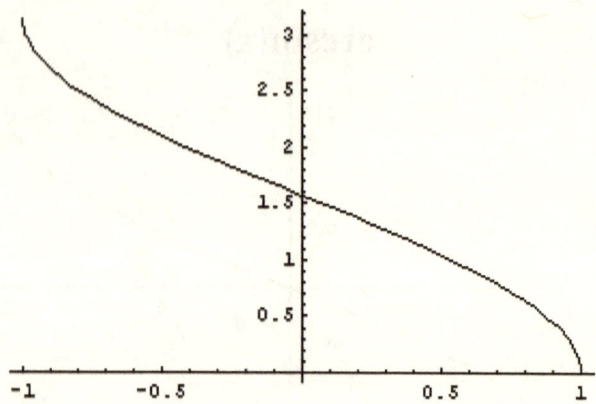

arcssec(x)

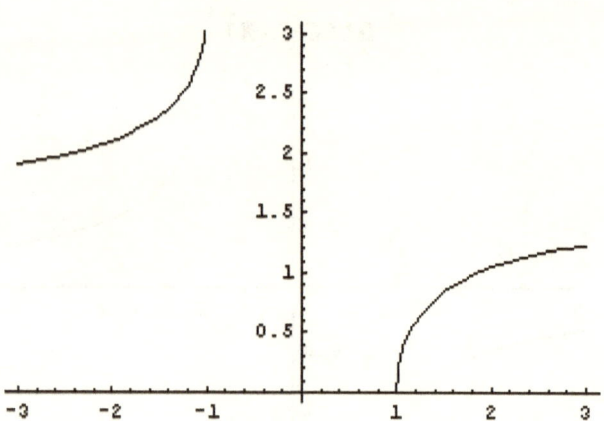

arctan(x)

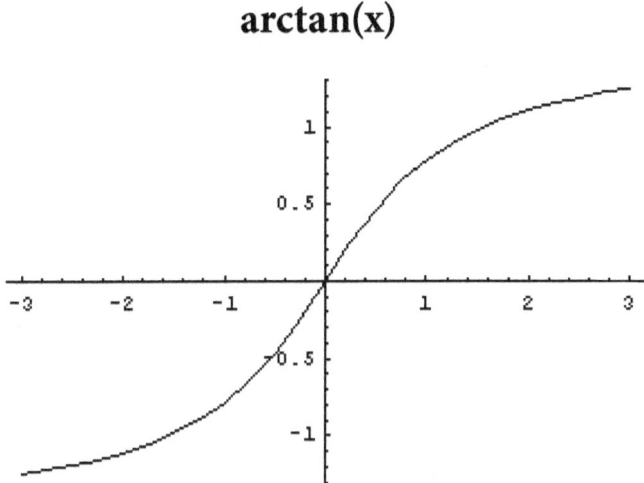

arccot(x)

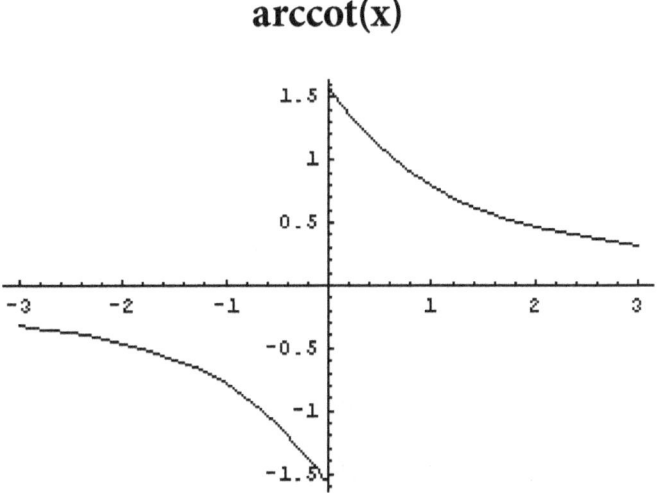

Basic Hyperbolic Function Definitions

$$\sinh(x) = \frac{e^x - e^x}{2}$$

$$\cosh(x) = \frac{e^x + e^x}{2}$$

$$\tanh(x) = \frac{e^x - e^x}{e^x + e^x}$$

$$\text{sech}(x) = \frac{2}{e^x + e^x}$$

$$\text{csch}(x) = \frac{2}{e^x - e^x}$$

$$\coth(x) = \frac{e^x + e^x}{e^x - e^x}$$

Relationships Between Hyperbolic Functions

$$\tanh(x) = \frac{\sinh(x)}{\cosh(x)}$$

$$\coth(x) = \frac{1}{\tanh(x)} = \frac{\sinh(x)}{\cosh(x)}$$

$$\text{sech}(x) = \frac{1}{\cosh(x)}$$

$$\text{csch}(x) = \frac{1}{\sinh(x)}$$

$$\cosh^2(x) - \sinh^2(x) = 1$$

$$\text{sech}^2(x) + \tanh^2(x) = 1$$

$$\coth^2(x) - \text{csch}^2(x) = 1$$

Basic Hyperbolic Function Graphs

sinh(x)

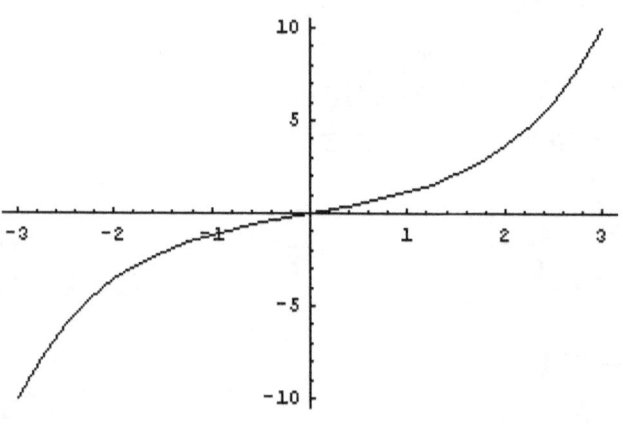

csch(x)

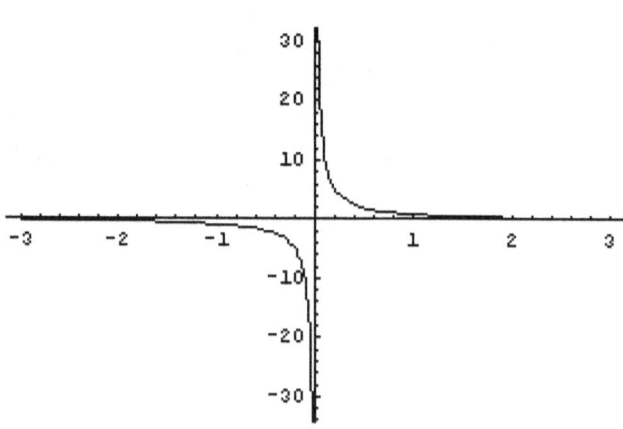

cosh(x)

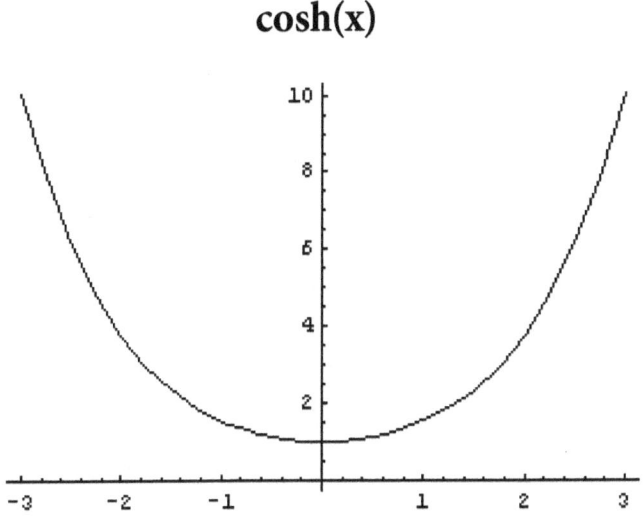

sech(x)

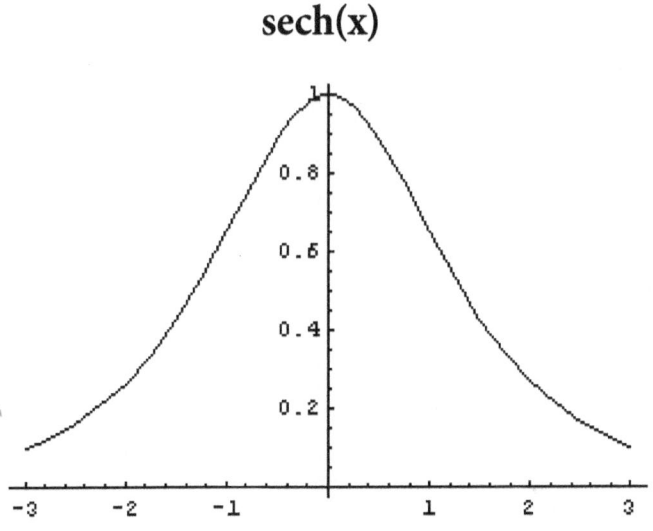

tanh(x)

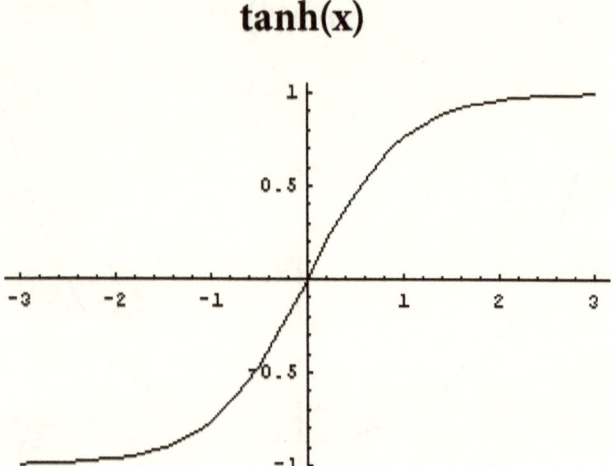

coth(x)

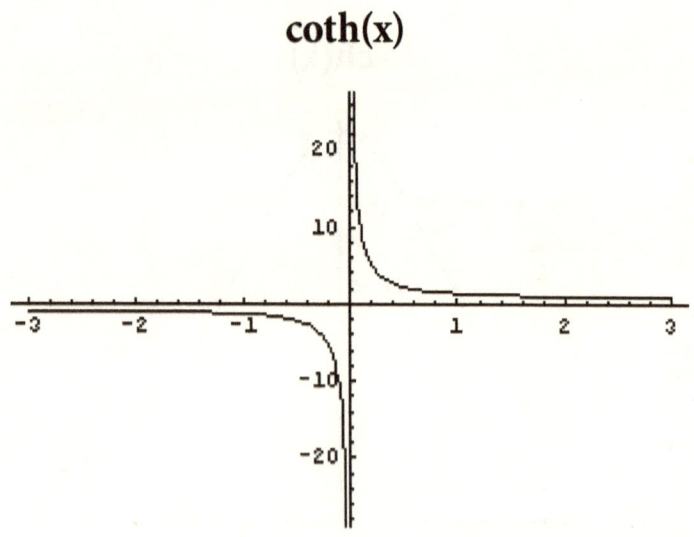

Basic Derivatives

$$\frac{d}{dx}[x^2] = 2x$$

$$\frac{d}{dx}[\sqrt{x}] = \frac{1}{2\sqrt{x}}$$

$$\frac{d}{dx}[\sin(x)] = \cos(x) \quad \text{and} \quad \frac{d}{dx}[\sin(2x)] = 2\cos(2x)$$

$$\frac{d}{dx}[\cos(x)] = -\sin(x) \quad \text{and} \quad \frac{d}{dx}[\cos(2x)] = -2\sin(2x)$$

$$\frac{d}{dx}[\tan(x)] = \sec^2(x)$$

$$\frac{d}{dx}[\cot(x)] = -\csc^2(x)$$

$$\frac{d}{dx}[\sec(x)] = \sec(x)\tan(x)$$

$$\frac{d}{dx}[\csc(x)] = -\csc(x)\cot(x)$$

$$\frac{d}{dx}[e^x] = e^x$$

$$\frac{d}{dx}[e^{2x}] = 2e^{2x}$$

$$\frac{d}{dx}[a^x] = \ln(x)\, a^x$$

$$\frac{d}{dx}[\ln(x)] = \frac{1}{x}$$

$$\frac{d}{dx}[\log_a(x)] = \frac{1}{(\ln(a))x}$$

$$\frac{d}{dx}[\arcsin(x)] = \frac{1}{\sqrt{1 - x^2}}$$

$$\frac{d}{dx}[\text{arccsc}(x)] = \frac{-1}{x\sqrt{x^2 - 1}}$$

$$\frac{d}{dx}[\arccos(x)] = \frac{-1}{\sqrt{1 - x^2}}$$

$$\frac{d}{dx}[\text{arcsec}(x)] = \frac{1}{x\sqrt{x^2 - 1}}$$

$$\frac{d}{dx}[\arctan(x)] = \frac{1}{1 + x^2}$$

$$\frac{d}{dx}[\text{arccot}(x)] = \frac{1}{1 + x^2}$$

Basic Integrals

$$\int x^2 \, dx = \frac{x^3}{3} + C$$

$$\int \sin(x) \, dx = -\cos(x) + C$$

$$\int \sin(2x) \, dx = -1/2 \cos(2x) + C$$

$$\int \cos(x) \, dx = \sin(x) + C$$

$$\int \sec^2(x) \, dx = \tan(x) + C$$

$$\int \csc^2(x) \, dx = -\cot(x) + C$$

$$\int \sec(x) \tan(x) \, dx = \sec(x) + C$$

$$\int \csc(x) \cot(x) \, dx = -\csc(x) + C$$

$$\int \tan(x) \, dx = \ln|\cos(x)| + C$$

$$\int \cot(x) \, dx = \ln|\sin(x)| + C$$

$$\int \sec(x) \, dx = \ln|\sec(x) + \tan(x)| + C$$

$$\int \csc(x)\,dx = \ln|\csc(x) + \cot(x)| + C$$

$$\int e^x\,dx = e^x + C$$

$$\int e^{2x}\,dx = \tfrac{1}{2} e^{2x} + C$$

$$\int 1/x\,dx = \ln|x| + C$$

$$\int a^x\,dx = \frac{a^x}{\ln|a|} + C \text{ when } a > 0, \text{ and } a \neq 1$$

NOTES:

NOTES:

NOTES:

NOTES:

NOTES:

NOTES:

NOTES:

NOTES:

www.ingramcontent.com/pod-product-compliance
Lightning Source LLC
Chambersburg PA
CBHW021045180526
45163CB00005B/2288